The ESS f

REAL VARIABLES

Tefera Worku, M.S.

Mathematics Instructor
State University of New York
Albany, New York

Research and Education Association
61 Ethel Road West
Piscataway, New Jersey 08854

THE ESSENTIALS® OF
REAL VARIABLES

Printed in the United States of America

Library of Congress Catalog Card Number 93-87627

International Standard Book Number 0–87891–921-X

ESSENTIALS is a registered trademark of
Research & Education Association, Piscataway, New Jersey 08854

WHAT "THE ESSENTIALS" WILL DO FOR YOU

This book is a review and study guide. It is comprehensive and it is concise.

It helps in preparing for exams, in doing homework, and remains a handy reference source at all times.

It condenses the vast amount of detail characteristic of the subject matter and summarizes the **essentials** of the field.

It will thus save hours of study and preparation time.

The book provides quick access to the important facts, principles, theorems, concepts, and equations in the field.

Materials needed for exams can be reviewed in summary form – eliminating the need to read and re-read many pages of textbook and class notes. The summaries will even tend to bring detail to mind that had been previously read or noted.

This "ESSENTIALS" book has been prepared by an expert in the field, and has been carefully reviewed to assure accuracy and maximum usefulness.

Dr. Max Fogiel
Program Director

Contents

CHAPTER 1

Set Theory and Topology of the Real Number System

1.1 Set Theory

A collection of numbers or points in the real number system viewed as a single entity is referred to as a set (denoted by capital letters, $A, B, C, D, \ldots$). These numbers or points are called elements or members of the set (denoted by lowercase letters, $a, b, c, d, \ldots$).

1.1.1 Equivalent Sets

1. Let A and B be two sets. A rule Γ, which associates with each element a of the set A exactly one element b of the set B, and under which each element b of B corresponds to exactly one element of A, is called a *one-to-one-correspondence* between set A and set B.

2. If it is possible to establish a one-to-one correspondence between two sets A and B, these sets are said to be *equivalent* or to have the same *cardinal number* (the term power is synonymous to cardinal number), and denoted as

$$A \sim B.$$

3. Example:

$\mathbf{O}$	1	3	5 ...	$2n + 1$ ...
$\mathbf{Z}_-$	-1	-2	-3 ...	$-n - 1$ ...

form a one-to-one correspondence between the set $\mathbf{O}$ of odd *natural numbers* and the set $\mathbf{Z}_-$ of *negative integers*.

1.1.2 Countability

1. Every set A equivalent to the set $\mathbf{N}$ of natural numbers is said to be *denumerable*.

2. A set which is finite or equivalent to $\mathbf{N}$ is referred to as *countable*.

3. Some examples of denumerable sets:

$$A = \{1, 4, 9, 16, \dots, n^2, \dots\}$$

$$\mathbf{Z} = \{\dots, -2, -1, 0, 1, 2, \dots\}$$

4. A set A is denumerable if and only if it is possible to enumerate it, i.e., to put it into the form of an ordinary infinite sequence:

$$A = \{a_1, a_2, a_3, \dots, a_n, \dots\}$$

5. Every infinite set A contains a denumerable subset D.

6. Every infinite subset of a denumerable set is denumerable.

7. The union of a countable number of denumerable sets is denumerable.

8. The set $\mathbf{Q}$ of all rational numbers is denumerable.

9. An infinite set which is not equivalent to the set $\mathbf{N}$ is called *non-denumerable*.

10. Example: The closed interval $[0, 1]$ is non-denumerable.

11. If a set A is equivalent to $B = [0, 1]$, i.e.,

$$A \sim B$$

then A is said to have the *power of the continuum*, or, more briefly, *the power of A is c*.

12. Every closed interval $[a, b]$, every open interval (a, b) has the power of c.

13. The set of all *irrational* numbers has power c.

1.1.3 Cantor Set

1. The cantor set $\mathbf{D}$ is a subset of $[0, 1]$ obtained by first removing the middle third

$$\left(\frac{1}{3}, \frac{2}{3}\right)$$

from $[0, 1]$, then removing the middle thirds

$$\left(\frac{1}{9}, \frac{2}{9}\right) \text{ and } \left(\frac{7}{9}, \frac{8}{9}\right)$$

from the remaining intervals, and continuing indefinitely.

2. $\mathbf{D}$ is compact, *nowhere dense* (i.e., its closure has empty interior), and has no isolated points (see definition in 1.2.1 below).

3. $\mathbf{D} = \left\{ \sum_{n=1}^{\infty} \frac{a_n}{3^n} : \quad a_n = 0 \text{ or } 2 \text{ for each } n \right\}$

and cardinality of $\mathbf{D} = c$.

1.1.4 Algebra of Sets

1. A collection $\mathcal{F}$ of subsets of a set $\mathbf{S}$ is called an *Algebra of Set* or a *Boolean Algebra* if

3

i) $E_1 \bigcup E_2$ is in $\mathcal{F}$ whenever E_1 and E_2 are;

ii) CE (complement of E) is in $\mathcal{F}$ whenever E is. Note if $\mathcal{F}$ is an algebra of sets E_1, $E_2 \in \mathcal{F}$, then:

    a) $CE_1 \in \mathcal{F}$ and $CE_2 \in \mathcal{F}$ by property (ii) above;

    b) $CE_1 \bigcup CE_2 \in \mathcal{F}$ by property (i) above;

    c) $CE_1 \bigcup CE_2 = C(E_1 \bigcap E_2)$ by De Morgan's law;

    d) $E_1 \bigcap E_2 \in \mathcal{F}$, since $E_1 \bigcap E_2 = C\big(C(E_1 \bigcap E_2)\big)$.

2. An algebra $\mathcal{F}$ of sets is called a σ-*algebra* (read as sigma algebra), or a *Borel Field*, if every union of a countable collection of sets in $\mathcal{F}$ is also in $\mathcal{F}$.

3. *Borel subsets* of **R**. The sets in the smallest σ-algebra containing the open subsets of **R** are called Borel sets (Borel subsets) of **R**. (The definition of the Borel sets in a metric space is similar.) Note that the Borel subsets of **R** are also in the smallest σ-algebra containing the closed and bounded intervals of **R**, and that they have cardinality c.

1.2 Point Sets

When using geometric terminology, we can refer to sets of real numbers as sets of points on the real line, sets of points in the plane, or sets of points in some higher-dimensional space. These points can be further categorized in terms of *limit points*, *closed sets*, and *open sets*.

1.2.1 Limit Points

1. The point X_0 is called a *limit point* of a point set E if every open interval containing this point contains at least one point of E distinct from the point X_0.

Note, the point X_0 itself may or may not belong to the set E.

Example: Even though the number 0 is the limit point for sets

$$E_1 = \left\{ 0, 1, \frac{1}{2}, \frac{1}{3}, \frac{1}{4}, \ldots \right\};$$

$$E_2 = \left\{ 1, \frac{1}{2}, \frac{1}{3}, \frac{1}{4}, \frac{1}{5}, \ldots \right\},$$

while it is the only limit point of E_1, it does not belong to E_2.

2. If the point X_0 belongs to the set E but is not a limit point of E, it is called an *isolated point* of the set E.

3. (Bolzano-Weirstrass Theorem) Every bounded infinite set E has at least one limit point (which may or may not belong to E).

4. Example: all elements of E_1 except 0 are isolated points of E_1 in the example above.

1.2.2 Closed Sets

1. The set of all limit points of E is called the *derived set* of the set E and is denoted by E'.

2. If $E' \subseteq E$, the set E is said to be *closed*.

3. If $E \subseteq E'$, the set E is said to be *dense in itself.*

4. If $E = E'$, the set E is said to be *perfect.*

5. The set $E \cup E'$ is called the *closure* of the set E and is denoted by $\overline{E}$.

Thus a set is closed if it contains all its limit points. A set which is dense in itself has no isolated points. A perfect set is one which is closed and dense in itself.

6. Examples:

1) The cantor set is a perfect set.

2) The set

$$E = \left\{ 1, \frac{1}{2}, \frac{1}{3}, \ldots, \frac{1}{n}, \ldots \right\}$$

is neither closed nor dense in itself $\left(\text{note } E' = \{0\} \right)$.

3) The set

$$E = \left\{ 1, \frac{1}{2}, \frac{1}{3}, \ldots, \frac{1}{n}, \ldots, 0 \right\}$$

is closed but is not dense in itself.

4) **Q** (the set of all rational numbers) is dense in itself but is not closed $\left(\text{note } \mathbf{Q}' = \mathbf{R} \right)$.

7. The *union* of a *finite* number of closed sets is closed.

8. The *intersection* of an *arbitrary* family of closed sets is closed.

9. Statement 7 is not necessarily true if finite is replaced by *infinite*:

$$\bigcup \left[\frac{1}{n}, 1 \right] = (0, 1)$$

which is not closed.

1.2.3 Open Sets

1. A point X_0 is called an *interior point* of the set E if there exists an open interval, contained entirely in the set E, which contains X_0:

$$X_0 \in (a, b) \subseteq E.$$

2. A set E is *open* if all of its points are interior points.

3. Open intervals: (a, b), $(-\infty, a)$, (b, ∞), the set of real number **R**, the empty set ϕ are all open sets (intervals).

4. The union of an arbitrary family of open sets is open.

5. The intersection of a finite collection of open sets is open.

    Remark: The intersection of an infinite number of open sets need not be an open set.

    $$\text{E.g. } \bigcap_{n=1}^{\infty}\left(-\frac{1}{n}, \ \frac{1}{n}\right) = \{0\}$$

    is *not* an open set.

6. If the set F is closed, then its complement is open.

7. If the set G is open, then its complement is closed.

CHAPTER 2

Metric Spaces

2.1 Definition and Basic Properties

The study of real variables is often dependent on a few properties of the distance between points and not on the fact that the points are in $\mathbf{R}$ or $\mathbf{R}^n$. When these properties of distance are investigated abstractly they lead to the concept of a metric space.

2.1.1 Metric Space d

A *metric* d on a non-empty set X is a function $d : X \times X \to \mathbf{R}$, which satisfies the following three conditions:

1. $d(x, y) \geq 0$, and $d(x, y) = 0 \Leftrightarrow x = y$;

2. $d(x, y) = d(y, x)$ (symmetry);

3. $d(x, y) \leq d(x, y) + d(z, y)$ (the triangle inequality).

2.1.2 Examples of a Metric

1. Let X be an arbitrary non-empty set, then

$$d(x, y) = \begin{cases} 0 \text{ if } x = y \\ 1 \text{ if } x \neq y \end{cases}$$

is a metric.

2. The function d defined on the real line **R** by $d(x, y) = |x - y|$ is a metric and is called the *usual metric* of **R**.

3. Let F be the set of all continuous (and bounded) real functions defined on the closed unit interval. The function d defined on $F \times F$ by

$$d(f,g) = \sup\left\{\left|f(x) - g(x)\right|\right\}$$
$$x \in [0,1]$$

is a metric.

2.1.3 A Metric Space

1. A *metric space* consists of two objects: a non-empty set X and a metric d on X. The elements of X are called the points of the metric space (X, d). Whenever it can be done without causing confusion, we denote the metric space (X, d) by the symbol X which is used for the underlying set of points.

2. Let X be a metric space with the metric d, and let A be a subset of X.

   If x is a point of X, then the *distance* from x to A is defined by

   $d(x, A) = \inf\{d(x, y) : y \in A\}$ (see illustration below)

   where if $x = [a, b]$ and $y = (c, d)$, then $d(x, y) = \sqrt{(a-c)^2 + (b-d)^2}$.

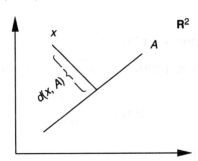

3. The *diameter* of the subset A of X is defined by

$$d(A) = \sup\{d(y_1, y_2) : y_1 \text{ and } y_2 \in A\}.$$

2.2 Open Sets

1. Let X be a metric space with metric d. If X_0 is a point of X and ε is a positive real number, the *open sphere* $S_\varepsilon (X_0)$ with center X_0 and radius ε is the subset of X defined by

$$S_\varepsilon(X_0) = \{x \in X : d(x_1, X_0) < \varepsilon\}.$$

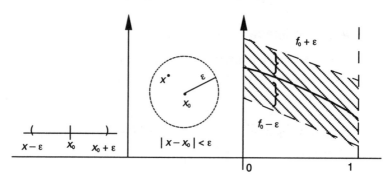

An open sphere on the real line.

An open sphere in the plane.

An open sphere on the set of all continuous functions on [0, 1].

2. A subset G of the metric space X is called an *open set* if, given any point x in G, there exists a positive real number ε such that $S_\varepsilon(x) \subseteq G$.

3. In any metric space X, the empty set ϕ and the full space X are open sets.

4. In any metric space X, each open sphere is an open set.

5. Let X be a metric space. A subset G of X is open $\Leftrightarrow$ if it is a union of open spheres.

6. Let X be a metric space. Then, a) any union of open sets in X is open; b) any finite intersection of open sets in X is open.

7. If X is the real line, then every non-empty open set in X is the union of a countable disjoint of open intervals.

2.3 Closed Sets

1. A subset A of X is closed if and only if its complement, CA is open.

2. In any metric space X, the empty set ϕ and the full space X are closed sets.

3. The notations and definitions in section 1.2.2 are the same as that used in the context of any topological space (see definition of topological space in section 2.5) or metric space.

2.4 Convergence, Completeness, and Baire's Theorem

With the definition of metric space, some important concepts can now be introduced.

2.4.1 Convergence

Let X be a metric space with metric d, and let

$$\{x_n\} = \{x_1, x_2, ..., x_n, ...\}$$

be a sequence of points in X. We say that $\{x_n\}$ is *convergent* if there exists a point x in X such that for each $\varepsilon > 0$, there exists a positive integer n_0 such that

$$\forall n \geq n_0 \Rightarrow d(x_n, x) < \varepsilon.$$

We usually symbolize this by writing

$$x_n \rightarrow x.$$

2.4.2 Completeness

1. A sequence $\{x_n\}$ is said to be *cauchy* if and only if for each $\varepsilon > 0$ there is n_0 such that

$$\forall m, \, n \geq n_0 \Rightarrow d\left(x_m, x_n\right) < \varepsilon.$$

2. A convergent sequence is cauchy.

3. A *complete* metric space is a metric space in which every cauchy sequence is convergent.

4. Let X be a complete metric space, and let Y be a subspace of X. Then Y is complete $\Leftrightarrow$ it is closed.

5. A sequence $\{A_n\}$ of subsets of a metric space is called a *decreasing sequence* if

$$A_1 \geq A_2 \geq A_3 \geq \ldots$$

6. Cantor's Intersection Theorem: Let X be a complete metric space, and let $\{F_n\}$ be a decreasing sequence of non-empty closed subsets of X such that $d(F_n) \rightarrow 0$, where $d(F_n) = \sup\{d(x, y) : x, y \, \varepsilon \, F_n\}$. Then

$$F = \bigcap_{n=1}^{\infty} F_n$$

contains exactly one point.

2.4.3 Baire's Theorem

1. If $\{A_n\}$ is a sequence of nowhere dense sets in a complete metric space X, then there exists a point in X which is not in

any of the A_n's; (A subset A of the metric space is said to be nowhere dense if its closure has empty interior).

2. If a compelte metric space is the union of a sequence of its subsets, then the closure of at least one set in the sequence must have non-empty interior.

2.4.4   Fixed Point Theorem

If T is a *contraction* defined on a complete metric space X (i.e., $T : X \rightarrow X$ such that $d(Tx, Ty) \leq \delta\, d(x, y)$ for some fixed $0 < \delta < 1$), then T has a unique fixed point, $x_f \, \varepsilon \, X$ (where $T(x_f) = x_f$).

Example:

$$\text{Let } X = \begin{bmatrix} 0, & 1 \end{bmatrix} \text{ with the metric } |\ .\ |, \text{ i.e.,}$$

$$d(x,y) = |x - y|.$$

$$\text{Let } T : [0,1] \rightarrow [0,1] \text{ be such that}$$

$$T(x) = \frac{1-x}{2}. \text{ Then}$$

$$d(Tx,\ Ty) = \left| \frac{1-x}{2} - \frac{1-y}{2} \right| = \left| \frac{x-y}{2} \right|.$$

And the unique fixed point is $x_f = \dfrac{1}{3}$.

2.5   Introduction to Topological Space

In order to study the properties of various functions in the product-space $\mathbf{R}^n$, one needs to know the concepts of topological space and mapping between two topological spaces. In addition, the concept of topological space also enables one to look at the metric space with a wider perspective.

2.5.1 Topological Space

1. Let X be a non-empty set. A class Ω of subsets of X is called a *topology* on X if it satisifes the following two conditions:

    i)   The union of every class of sets in Ω is in Ω;

    ii) The intersection of finite class of sets in Ω is a set in Ω.

2. A *topological space* consists of two objects: a non-empty set X and a topology Ω on X.

3. The sets in the class Ω in 2) above are called the *open sets* of the topological space (X, Ω) and the elements of X are called its points.

2.5.2 Continuity

1. Let X,Y be topological spaces and f a mapping of X into Y. f is called a *continuous mapping* if $f^{-1}(G)$ is open in X whenever G is open in Y, and *open mapping* if $f(G)$ is open in Y whenever G is open in X.

2. A *homeomorphism* is a one-to-one continuous mapping of one topological space onto another, which is also an open mapping.

2.5.3 Metric Space as a Topological Space

Let X be a metric space, and let the topology be the class of all subsets of X which are open in the sense of definition in section 2.2. This is called the *usual topology* on a metric space, and we say that these sets are the open sets *generated* by the metric on the space.

2.5.4 Separability

1. A subset A of a topological space X is said to be *dense* (or everywhere dense) if $\overline{A} = X$.

2. A topological space X is called a *separable space* if it has a countable dense subset. For instance, **R**, i.e., the set of real numbers, since Q (the set of rational numbers) is dense in **R** and **Q** is countable.

CHAPTER 3

Measure Theory

3.1 Measure on an Algebra

In this section we will define precisely what a measure of a set is and also present some basic properties of measures. And we will do so with respect to an algebra.

3.1.1 Definition

Let $\mathcal{A}$ be an algebra of subsets of X and μ be an extended real-valued function (a function which may possibly take values $-\infty$, ∞) on $\mathcal{A}$. Then μ is called a *measure* on $\mathcal{A}$ if

i) $\mu(\phi) = 0$:

ii) $\mu(A) \geq 0$ if $A \in \mathcal{A}$:

iii) $\mu\left(\bigcup_{n=1}^{\infty} A_n\right) = \sum_{n=1}^{\infty} \mu(A_n)$

for every sequence A_n of pairwise disjoint (i.e. $A_n \cap A_m = \phi$ is $n \neq m$) sets of $\mathcal{A}$ with

$$\bigcup_{n=1}^{\infty} A_n \in \mathcal{A}$$

Above, iii) is known as the *countable additivity* property of μ. μ is called a *finitely additive measure* if iii) is replaced by

iv) $\quad \mu\left(\bigcup_{n=1}^{K} A_n\right) = \sum_{n=1}^{K} \mu(A_n),$

for every finite sequence

$$\left\{A_n\right\}_{n=1}^{K}$$

of pairwise disjoint sets in $\mathcal{A}$.

An example of measure:

Let $\mathcal{A} = 2^x$ (the class of all subsets of a set X).

Let μ be defined on $\mathcal{A}$ by

$$\mu(A) = \begin{cases} \text{number of points in } A, & \text{if } A \text{ is finite;} \\ \infty, & \text{if } A \text{ is infinite.} \end{cases}$$

Then μ is a measure. This measure is known as the *counting measure* in X.

3.1.2 Properties of Measure

Let μ be a measure on an algebra $\mathcal{A}$. Then

1. $A \subseteq B,\ A \in \mathcal{A},\ B \in \mathcal{A}$ implies $\mu(A) \leq \mu(B)$

2. $\mu\left(\bigcup_{n=1}^{\infty} A_n\right) \leq \sum_{n=1}^{\infty} \mu(A_n)$(the countable subadditive property),

    if $A_n \in \mathcal{A},\ 1 \leq n < \infty,\ A_n \cap A_m = \phi$ for $n \neq m$, and $\bigcup_{n=1}^{\infty} A_n \in \mathcal{A}$.

3. If $A_n \supseteq A_n + 1,\ A_n \in \mathcal{A}$, for $1 \leq n \leq \infty$. $\mu(A_1) < \infty$ and

$$\bigcap_{n=1}^{\infty} A_n \in \mathcal{A}, \text{ then}$$

$$\mu\left(\bigcap_{n=1}^{\infty} A_n\right) = \lim_{n \to \infty} \mu(A_n)$$

4. If $A_n \subseteq A_n + 1$, $A_n \in \mathcal{A}$ for $1 \leq n < \infty$ and

$$\bigcup_{n=1}^{\infty} A_n \in \mathcal{A}, \text{ then}$$

$$\mu\left(\bigcap_{n=1}^{\infty} A_n\right) = \lim_{n \to \infty} \mu(A_n)$$

3.1.3 Finite and σ-Finite Measure

A measure μ on an algebra $\mathcal{A}$ of subsets of X is called *finite* if $\mu(x) < \infty$. It is called σ-*finite* if there is a sequence $\{X_n\}$ of sets in $\mathcal{A}$ with $\mu(X_n) < \infty$ and

$$X = \bigcup_{n=1}^{\infty} X_n$$

3.2 Lebesgue Measure

Having defined what a measure of an algebra is in the last section, this section applies it to the simple sets in **R**, intervals. The resultant measure is the Lebesgue measure on intervals.

3.2.1 Algebra of Intervals

Let X be **R** and $\mathcal{A}$ be the class of all finite disjoint unions of right-closed, left-open intervals including intervals of the form $(-\infty, a]$, (a, ∞) and $(-\infty, \infty)$, $-\infty < a < \infty$, and the empty set. Then $\mathcal{A}$ is an algebra.

3.2.2 Measure on Algebra of Intervals

For $A \in \mathcal{A}$, let

$$M_o = \sum_{n=1}^{K} l(I_n), \text{ if } A = \bigcup_{n=1}^{K} I_n.$$

Then M_o is well defined, and for B, $C \in \mathcal{A}$ and $B \subseteq C$, $M_o(B) \leq M_o(C)$. ($l(I_n)$ is the length of an interval I_n which is infinite if I_n is an infinite interval.)

M_o is a measure on $\mathcal{A}$ and is called the *Lebesgue measure* on intervals.

3.3 Construction of Measures

Measurability, complete measure, Carathéodory extension theorem, and measurable space are the topics for this section. These topics are meant to help one with the construction of measures.

3.3.1 Outer Measure

1. By an outer measure μ^* we mean an extended real valued set function defined on 2^x, having the following properties:

   i) $\mu^*(\phi) = 0$

   ii) $\mu^*(A) \leq \mu^*(B)$, if $A \subset B \subset X$

   iii) If $\{E_n\}$ is a countable family of subsets of X, then

   $$\mu^*\left(\bigcup_{n=1}^{\infty} E_n\right) \leq \sum_{n=1}^{\infty} \mu^*(E_n).$$

2. Example:

   suppose $\mu^*(E) = \begin{cases} 0, & E = \phi \\ 1, & E \neq \phi \end{cases}$

Then μ^* is an outer measure, which is not a measure on 2^x, if x has at least two points.

3. Let $\mathcal{F}$ be a class of subsets X containing the empty set such that for every $A \subset X$, there exists a sequence

$$\{B_n\}_{n=1}^{\infty}$$

from $\mathcal{F}$ such that

$$A \subset \bigcup_{n=1}^{\infty} B_n.$$

Let τ be an extended real-valued function on $\mathcal{F}$ such that $\tau(\phi) = 0$ and $\tau(A) \geq 0$ for $A \in \mathcal{F}$. Then μ^* is defined on 2^x by

$$\mu^*(A) = \inf\left\{\sum_{n=1}^{\infty} \tau(B_n) : B_n \in \mathcal{F}, A \subset \bigcup_{n=1}^{\infty} B_n\right\}$$

is an outer measure.

4. Example:

$$\text{Let } X = \mathbf{N} = \{1, 2, 3, \ldots\}$$

$$\mathcal{F} = \{(n) : n = 1, 2, \ldots\} \cup \{\phi\}$$

suppose

$$\tau(E) = \begin{cases} 0, & E = \phi \\ 1, & E \neq \phi \end{cases}$$

then

$$\mu^*(A) = \begin{cases} \infty, & \text{if } A \text{ is finite} \\ \text{the number of points in } A, & \text{if } A \text{ is finite.} \end{cases}$$

3.3.2 Lebesgue Outer Measure

Let $X = \mathbf{R}$ and $\mathcal{F}$ be the class of all left open, right-closed intervals (including finite or infinite of the form $(-\infty, a]$, $(a, b]$, or (a, ∞) and the empty set.

Let $\tau(\phi) = 0$, and $\tau(I) = l(I)$, i.e., length of I, for every interval $I \in \mathcal{F}$. Then the outer measure μ^*, induced by τ, on $2^{\mathbf{R}}$ (the class of *all* subsets of $\mathbf{R}$) is called the *Lebesgue outer measure* on $\mathbf{R}$ defined as

$$\mu^*(A) = \inf\left\{ \sum_{n=1}^{\infty} l(I_n) : A \subset \bigcup_{n=1}^{\infty} I_n,\ I_n \in \mathcal{F} \right\}.$$

3.3.3 Measurability

1. $E \subset X$ is called μ^*-measurable if for every $A \subseteq X$, $\mu^*(A) = \mu^*(A \cap E) + \mu^*(A \cap E^c)$ (where $E^c = CE$ is the complement of E).

2. The class $\mathcal{B}$ of μ^*-measurable sets is a σ-algebra. Also $\overline{\mu}$ the restriction of μ^* to $\mathcal{B}$ (μ^* acting on smaller subcollection $\mathcal{B}$), is a measure.

3.3.4 A Complete Measure

1. $\mu^*(A) = 0$ implies A is μ^*-measurable;

2. due to statement 1 the measure $\overline{\mu}$, (the restriction of μ^* on $\mathcal{B}$) has the following property:

    $E \in \mathcal{B}$, $\overline{\mu}(E) = 0$, and $F \subset E$ implies $F \in \mathcal{B}$;

3. A measure having the property in 2. is called a *complete measure*.

3.3.5 Carathéodory Extention Theorem

Let μ be a measure on an algebra $A \subset 2^x$. Suppose $E \subseteq X$, and

$$\mu^*(E) = \inf\left\{\sum_{i=1}^{\infty} \mu(E_i) : E \subseteq \bigcup_{i=1}^{\infty} E_i, \ E_i \in \mathcal{A}\right\}.$$

Then the following properties hold:

a) μ^* is an outer measure.

b) $E \in \mathcal{A}$ implies $\mu(E) = \mu^*(E)$

c) $E \in \mathcal{A}$ implies E is μ^*-measurable.

d) The restriction $\overline{\mu}$ of μ^* to the μ^*-measurable sets is an extension of μ to a measure on a σ-algebra containing $\mathcal{A}$.

e) If μ is σ-finite, then $\overline{\mu}$ is the only measure (on the smallest σ-algebra containing $\mathcal{A}$) that is an extension of μ.

3.3.6 Remarks on Lebesgue Measure

1. Since the Lebesgue outer measure defined in section 3.3.2 is the most important outer measure on **R**, it is commonly denoted by m^*.

2. $m^*(A) = \inf\left\{\sum_{n=1}^{\infty} l(I_n) : A \subseteq \bigcup_{n=1}^{\infty} I_n\right\},$

   where I_n's are intervals of the form $[-\infty, a]$, $[a, b]$, or $[a, \infty)$.

3. m^* restricted to m^*-measurable sets called the Lebesgue measurable sets and denoted by $\mathcal{M}$ is a measure. This is the well-known Lebesgue measure on **R**. We denote it by m.

4. $\mathcal{M}$ contains all Borel sets, but *not* all elements of $\mathcal{M}$ are Borel sets.

5. m is translation invariant, i.e., for each Lebesgue measurable set $A \subseteq \mathbf{R}$ and $X \in \mathbf{R}$, $A + X$ is Lebesgue measurable and $m(A + X) = m(A)$.

6. Not every subset of $\mathbf{R}$ is Lebesgue measurable. In fact, let a relation $\sim$ defined in $\mathbf{R}$ by:

$$x \sim y \Leftrightarrow x - y \text{ is a rational number.}$$

Then $\sim$ is an equivalence relation and partitions $[0, 1)$ into a set of equivalent classes. By *axiom of choice* there is a set S which contains exactly one element from each equivalent class. It turns out that S is not Lebesgue measurable. The equivalence relation and equivalence classes have the usual definition, while the axiom of choice states that given any non-empty class of non-empty sets, a set can be formed which contains precisely one element taken from each set in the given class.

3.3.7 Measure Space

1. A triple $(X, \mathcal{A}, \mu)$ is called a *measure space* if

   i) X is a non-empty set;

   ii) $\mathcal{A}$ is a σ-algebra of subsets of X, and

   iii) μ is a measure of set $\mathcal{A}$.

2. In statement 1 when X is $\mathbf{R}$, $\mathcal{A}$ is the class of Lebesgue-measurable sets, and μ is the Lebesgue measure, then the measure space is referred to as the *Lebesgue measure space*.

3. In statement 1 if $\mathcal{A}$ is the class of Borel sets of $\mathbf{R}$, the measure space is called the *Borel measure space*.

CHAPTER 4

Measurable Function

4.1   Definition

Let $E \in \mathcal{A}$ and $\mathbf{R}_\infty$ be the extended reals, i.e.,

$$\mathbf{R}_\infty = \mathbf{R} \cup \{\infty\}.$$

A function $f : E \to \mathbf{R}_\infty$ is called *measurable* if for each α, the set $\{x \in E : f(x) > \alpha\} \in \mathcal{A}$. If $\mathcal{A}$ is the class of Lebesgue-measurable subsets (respectively Borel sets) on $\mathbf{R}$ (=X), a measurable function f is usually called a *Lebesgue* (respectively Borel) *measurable function*.

4.2   Characterizations of Measurable Functions

Let f be as defined above. Then the following are equivalent:

i) f is measurable.

ii) $\{x \in E : f(x) \geq \alpha\} \in \mathcal{A}$, if $\alpha \in \mathbf{R}$.

iii) $\{x \in E : f(x) < \alpha\} \in \mathcal{A}$, if $\alpha \in \mathbf{R}$.

iv) $\{x \in E : f(x) \leq \alpha\} \in \mathcal{A}$, if $\alpha \in \mathbf{R}$.

Moreover, these statements imply

v) $\{x \in E : f(x) = \alpha\} \in \mathcal{A}$, for each extended real number α.

4.3 Examples

1. If $X = \mathbf{R}$ and $\mathcal{A}$ is the class of Lebesgue-measurable sets or Borel sets on $\mathbf{R}$, then *every continuous function f on $E \in \mathcal{A}$ is measurable* since $\{x \in E : f(x) > \alpha\}$ is the intersection of E and an open set.

2. The characteristic function $\chi_A(x)$ is a measurable function on X if and only if $A \in \mathcal{A}$ (χ_A is defined by $\chi_A(x) = 1$ if $x \in A$ and $\chi_A = 0$ if $x \notin A$).

4.4 Properties of Measurable Function

Having learned what a measurable function is one needs to know the various properties of it. The following are important properties of measurable functions.

4.4.1 Functions of Measurable Functions

1. If f and g are measurable real-valued functions having the same domain, then $f \pm g$, $|f|$, $f \vee g$, $f \wedge g$, $f \times g$ are all measurable functions.

    Note: $f \vee g(x) = \max\{f(x),\ g(x)\}$ and
    $$f \wedge g(x) = \min\{f(x),\ g(x)\}.$$

2. If $\{f_n\}$ is a sequence of measurable functions (having the same domain), then

    $$\sup_n(f_n),\ \inf_n(f_n),\ \inf_n \sup_{K \geq n}(f_k),\ \sup_n \inf_{K \geq n}(f_n)$$

    are all measurable.

3. If f is a measurable function on $[a, h]$, and $\varepsilon > 0$, then there exists a continuous function g such that $m\{x \in [a, h] : |f(x) - g(x)| \geq \varepsilon\} < \varepsilon$.

4.4.2   Relation to Simple Functions

1. A real-valued function φ is called *simple* if it is measurable and assumes only a finite number of values.

2. A simple function has the following properties:

    i) If φ is simple and has the values $c_1, \ldots, c_n$, then

    $$\varphi = \sum_{i=1}^{n} c_i \, \chi_{Ai},$$

    where $A_i = \{x : \varphi(x) = c_i\}$

    ii) The *sum, product,* and *difference* of two simple functions are simple.

    iii) Let f be a measurable function. Then f is the pointwise limit of a sequence of simple functions. If f is bounded, then the convergence is uniform. If $f \geq 0$, then the above sequence can be taken as monotonic increasing.

4.4.3   Egroff's Theorem

If $\{f_n\}$ is a sequence of measurable functions which converge to a real-valued function f a.e. (a.e. means *"almost everywhere"*; a property holds a.e. if the set of points for which it fails to hold is measurable and has measure zero) on a measurable set E of finite measure, then given $\delta > 0$, there is a subset $F \subset E$ with $m(F) < \delta$ such that f_n converges to f uniformly on $E \sim F$.

4.4.4   Convergence in Measure

Definition: A sequence $\{f_n\}$ of measurable functions is said to converge in measure if given $\varepsilon > 0$, there is an N such that for all $n \geq N$ we have

$$m\{x : |f(x) - f_n(x)| \geq \varepsilon\} < \varepsilon.$$

Remarks:

1. Almost everywhere convergence implies convergence in measure, when $\mu(x) < \infty$. This follows from Egoroff's theorem. But

2. Convergence in measure doesn't necessarily imply convergence a.e. For instance, take the sequence of measurable functions $\{f_n\}$, where

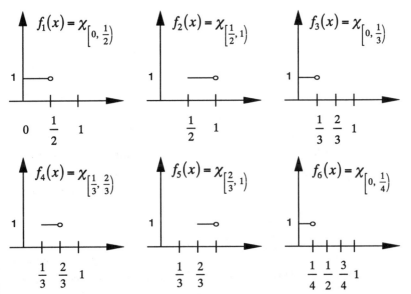

Similarly,

$$f_7(x) = \chi_{[\frac{1}{4}, \frac{1}{2}]}, \ f_8(x) = \chi_{[\frac{1}{2}, \frac{3}{4}]}, \text{ and so on.}$$

Thus, this sequence converges in measure to $f \equiv 0$.

But for every $x \in [0, 1]$ there is a subsequence of numbers $\{f_{n_j}(x)\}$ that converges to 1. Hence f_n does not converge to 0 pointwise.

3. If $\mu(x) = \infty$, then almost everywhere convergence does not necessarily imply convergence in measure. For instance,

    let $X = \{1, 2, 3, \ldots, n, \ldots\}$ with counting measure,

    and $f_n(x) = \chi_{\{n\}}$.

    then $f_n(x) \to 0$ for all $x \in X$. But if

    $$\in = \frac{1}{2},$$

    then

    $$m\left\{x: |f_n(x)| \geq \frac{1}{2}\right\} = \infty.$$

4. Although convergence in measure (by remarks in section 1.2) is more general than convergence a.e., the following theorem nevertheless holds:

    Theorem (F. Riesz): Let $\{f_n(x)\}$ be a sequence of functions which converges in measure to the function $f(x)$. Then there exists a subsequence

    $$f_{n1}(x), \ f_{n2}(x), \ f_{n3}(x), \ \ldots \ (n_1 < n_2 < n_3 < \ldots)$$

    which converges to the function $f(x)$ a.e.

    From the example in statement 2 above,

    $$\{f_2(x), \ f_5(x), \ f_9(x), \ f_{14}(x), \ f_{20}(x), \ f_{27}(x), \ \ldots \ \}$$

    is one such subsequence, i.e.,

    $$\left\{\chi_{[\frac{1}{2},1]}, \ \chi_{[\frac{2}{3},1]}, \ \chi_{[\frac{3}{4},1]}, \ \ldots, \ \chi_{[\frac{k-1}{k},1]}, \ \ldots \right\}.$$

4.4.5 Further Properties of Measurable Functions

1. Every function defined on a set of measure zero is measurable.

2. Given two functions f and g, if f is measurable and $f \equiv g$ a.e., then g is measurable.

3. Two functions $f(x)$ and $g(x)$ defined on the same set E are said to be equivalent if $m(\{x : f(x) \neq g(x)\}) = 0$. It is common to indicate that functions $f(x)$ and $g(x)$ are equivalent by

$$f(x) \sim g(x).$$

4. If the function $f(x)$ defined on the set E is measurable, then the function f^2 is measurable. (Since

$$\frac{1}{4}\left[(f+g)^2 - (f-g)^2\right] = f \cdot g$$

$f \cdot g$ is measurable if f and g are measurable.)

4.5 Approximation Theorems

1. Bernstein Polynomial: Let $f(x)$ be a finite function defined on the closed interval $[0, 1]$. The polynomial

$$B_n(x) = \sum_{K=0}^{n} f\left(\frac{k}{n}\right)\binom{n}{k}x^k(1-x)^{n-k},$$

is called the *Bernstein Polynomial* of degree n for the function $f(x)$.

Theorem (S. N. Bernstein): If the function $f(x)$ is continuous on the interval $[0, 1]$, then

$$B_n(x) \rightarrow f(x)$$

uniformly with respect to x, as $n \rightarrow \infty$.

2. Theorem (Weierstrass): If f is a continuous complex function on $[a, h]$, there exists a sequence of polynomials P_n such that

$$\lim_{n \to \infty} P_n(x) = f(x)$$

uniformly on $[a, h]$. If f is real, then P_n may be taken real.

3. Theorem (M. Fréchet): For every measurable function $f(x)$ which is defined on $[a, b]$ and is finite a.e., there exists a sequence of polynomials converging to $f(x)$ a. e.

CHAPTER 5

The Lebesgue Integral

5.1 Definitions

1. A *simple* function f is said to be *integrable* if it can be written as

$$\sum_{i=1}^{n} c_i \chi_{Ai}$$

such that $m(A_i) < \infty$ ($m(A_i)$ stands for Lebesgue measure of A_i) whenever $c_i \neq 0$. If f is integrable or non-negative, we write

$$\int f du = \sum_{i=1}^{n} C_i \, m(A_i).$$

If E is a Lebesgue measurable set, we write

$$\int_E f du = \int f \cdot \chi_E du.$$

2. Let f be a *non-negative* measurable function. Let E be a measurable set and E is a subset of the domain of f. Then we define

$$\int_E f du = \sup\left\{\int_E g du : 0 \le g \le f,\ g \text{ is a simple function}\right\}$$

or

$$\int_E f du = \inf\left\{\int_E g du : g \ge f,\ \text{is a simple function}\right\}.$$

If

$$\int_E f du < \infty,$$

then f is called *Lebesgue integrable* on E.

3. Let f be a measurable function whose domain contains a measurable set E. Let

$$f^+ = \frac{|f| + f}{2}$$

and

$$f^- = \frac{|f| - f}{2}.$$

f is *integrable* on E if and only if f^+ and f^- are both integrable on E.

5.2 Properties of Lebesgue Integral

Let f and g be non-negative measurable functions whose domains contain a measurable set E. Then

1. If $f \le g$ a.e., then

$$\int_E f du \le \int_E g du.$$

2. If $f = g$ a.e., then

$$\int_E f du = \int_E g du.$$

3. If

$$\int_E f du = 0,$$

then $f = 0$ a.e. on E.

4. If

$$\int_E f du \le \int_E g du$$

for all measurable set E, then $f \le g$ a.e.

5. If

$$\int_E f du < \infty,$$

then $\mu(A) = 0$, where $A = \{x \in E : f(x) = \infty\}$. The reason is that for each positive integer n, $0 \le n \cdot \chi_A \le f$ and therefore

$$n\mu(A) \le \int_E f du.$$

6. If $\mu(E) < \infty$ and $m \le f \le M$, m and M being two non-negative real numbers, then

$$m\mu(E) \le \int_E f du \le M\mu(E).$$

If f and g are integrable on a measurable set E, and a and b are real numbers, then we have a linear integral

$$\int_E (af + bg) du = a\int_E f du + b\int_E g du.$$

5.3 Passage to the Limit Under the Integral Sign

For a convergent sequence of measurable functions, its integral may also be convergent. The following theorems provide some general conclusions on this subject.

5.3.1 The Dominated Convergence Theorem (also known as the Lebesgue Convergence Theorem)

If $\{f_n\}$ is a sequence of measurable functions converging a.e. to a measurable function f such that $|f_n| \le g$ a.e. on a measurable set E, where g is an integrable function on E, then

$$\int_E f\,du = \lim_{n \to \infty} \int_E f_n\,du$$

5.3.2 Generalization of the Dominated Convergence Theorem

Let $\{f_n\}$ and $\{g_n\}$ be two sequences of measurable functions which converge a.e. to the measurable functions f and g, respectively. Suppose $|f_n| \le g_n$ and

$$\lim_{n \to \infty} \int_E g_n\,du = \int_E g\,du < \infty.$$

Then

$$\lim_{n \to \infty} \int_E f_n\,du = \int_E f\,du.$$

5.3.3 The Monotone Convergence Theorem

If $\{f_n\}$ is a non-decreasing sequence of non-negative measurable functions converging to a measurable function f a.e., then for any measurable set E,

$$\int_E f \, du = \lim_{n \to \infty} \int_E f_n \, du.$$

5.3.4   Fatou's Lemma

If $\{f_n\}$ is a sequence of non-negative measurable functions whose domains contain a measurable set E, then

$$\int_E \left(\lim_{n \to \infty} f_n \right) du \leq \lim_{n \to \infty} \int_E f_n \, du.$$

$$\left(\text{where } \lim_{n \to \infty} x_n = \sup_n \left(\inf_{m \geq n} x_m \right) \right)$$

5.4   Some Consequences of the Above Theorems

1.   If f is integrable on a measurable set E and if

$$E = \bigcup_{n=1}^{\infty} E_n, \ E_n \cap E_m = \phi \text{ for } n \neq m,$$

and E_n is measurable for every n, then

$$\int_E f \, du = \sum_{n=1}^{\infty} \int_{E_n} f \, du.$$

(This is a consequence of the Monotone Convergence Theorem.)

2.   If f is integrable on a measurable set E, then $A = \{x \in E : f(x) \neq 0\}$ has a σ-finite measure, i.e.,

$$A = \bigcup_{n=1}^{\infty} A_n,$$

such that $\mu(A_n) < \infty$ for every n. (This is a consequence of the Lebesgue Convergence Theorem.)

3. If f and g are integrable on a measurable set E, and $f \leq g$ a. e., then

$$\int_E f du \leq \int_E g du.$$

This is not a direct consequence of any of the theorems. It follows from the fact that if $f \leq g$ a. e., then

$$f^+ + g^- \leq g^+ + f^-$$

and property 2 in section 5.2.

5.5 Product Measures

Let $(X, \mathcal{A}, \mu)$ and $(Y, \mathcal{B}, \nu)$ be two measure spaces and $X \times Y$ be the cartesian product of X and Y. If $A \leq X, B \leq Y$, then $A \times B$ is called a *rectangle* (a *measurable rectangle*) when $A \in \mathcal{A}$ and $B \in \mathcal{B}$.

Now, let $\mathcal{R} = \{A \times B / A \in \mathcal{A} \text{ and } B \in \mathcal{B}\}$. $\mathcal{R}$ has the following properties.

1. The function λ defined on $\mathcal{R}$ by $\lambda(A \times B) = \mu(A) \times \nu(B)$ can be extended to a complete measure on the σ-algebra generated by $\mathcal{R}$ in some cases: for example $X = Y = \mathbf{R}$ and $\mu = \nu =$ Lebesgue measure on $\mathbf{R}$.

2. If $E \in \mathcal{A} \times \mathcal{B}$, i.e., $\mathcal{R}$, then

   i) $E_x \in \mathcal{B}, E_x = \{y \in Y : (x, y) \in E\}$;

   ii) $E_y \in \mathcal{A}, E_y = \{x \in X : (x, y) \in E\}$.

3. Let f be an extended real-valued $\mathcal{A} \times \mathcal{B}$ measurable (i.e. f^{-1} $(a \in \mathbf{R} \mid a > \alpha)$ is in the σ-algebra generated by $\mathcal{R}$, for every α) function on $X \times Y$. Then

    i) fx, where $fx(y) = f(x, y)$ is $\mathcal{B}$-measurable for all $x \in X$, and

    ii) f^y where $f^y(x) = f(x, y)$, is $\mathcal{A}$-measurable for all $y \in Y$.

5.6   Fubini's Theorem

Let μ and v be complete (refer to section 3.3.4 for the definition of complete measure) and f be integrable with respect to $\mu \times v$ on $\mathcal{M}$ (where $\mathcal{M}$ is the σ-algebra of λ-measurable sets mentioned in property 1 of section 5.5, and $\mu \times v$ is a commonly used notation for λ). Then

   i) $\int_Y f(x,y)\, dv(y)$ and $\int_X f(x,y)\, d\mu(x)$

       are integrable functions of X and Y respectively, and

   ii) $\int_{X \times Y} f\, d(\mu \times v) = \int_Y \left[\int_X f\, d\mu \right] dv = \int_X \left[\int_Y f\, dv \right] d\mu.$

5.7   Tonelli's Theorem

Let μ and v be complete and σ-finite. Let f be a *non-negative* $\mathcal{M}$-measurable function on $X \times Y$. Then

   i) $\int_Y f(x, y)\, dv(y)$ and $\int_X f(x, y)\, d\mu(x)$

       are integrable functions of X and Y respectively, and

   ii) $\int_{X \times Y} f\, d(\mu \times v) = \int_Y \left[\int_X f\, d\mu \right] dv = \int_X \left[\int_Y f\, dv \right] d\mu.$

   iii) For almost all x, $f_x^{(y)} = f(x, y)$ is $\mathcal{B}$-measurable, and for almost all y, $f^y(x) = f(x, y)$ is $\mathcal{A}$-measurable.

iv) $\int_Y f(x, y) \, dv(y)$ and $\int_X f(x, y) \, d\mu(x)$

are both measurable functions of x and y, respectively.

v) $\int_{X \times Y} f \, d(\mu \times v) = \int_Y \left[\int_X f \, d\mu \right] dv = \int_X \left[\int_Y f \, dv \right] d\mu.$

5.8 Remarks and Examples

1. If you let $X = Y = \mathbf{R} =$ the set of real numbers, and $\mu = v = m$ = Lebesgue measure, then, $\mu \times v$ is the Lebesgue measure on $\mathbf{R}^2$.

2. The integrability condition in Fubini's Theorem and the non-negativeness (as well as σ-finiteness) condition in Tonnelli's Theorem are essential, as the following examples show.

Examples:

1. Let $X = Y = (0, 1)$, $\mu = v =$ Lebesgue measure;

$$f(x,y) = \frac{x^2 - y^2}{x^2 + y^2}, \quad (x, y) \in (0, 1) \times (0, 1).$$

Then,

$$\int_0^1 f(x, y) dy = \frac{1}{1 + x^2}$$

and therefore,

$$\int_0^1 \left[\int_0^1 f(x, y) dy \right] dx = \frac{\pi}{4}.$$

But

$$\int_0^1 \left[\int_0^1 f(x, y) dx \right] dy = -\frac{\pi}{4}.$$

Notice if for $0 < x < 1$

$$\int_0^1 f(x, y)\, dy = \frac{1}{2x}.$$

It follows that

$$\int_0^1 |f(x, y)|\, dy \geq \frac{1}{2x}$$

and

$$\int_0^1 \left[\int_0^1 |f(x, y)|\, dy \right] dx \geq \frac{1}{2} \int_0^1 \frac{dx}{x} = \infty,$$

which means that f is not Lebesgue integrable on $(0, 1) \times (0, 1)$.

2. Let $I = [0, 1] \times [0, 1]$ and μ_1, μ_2 be the Lebesgue and the counting measures, respectively on $[0, 1]$.

Let $f(x, y) = X_\Delta(x, y)$, where

$$\Delta = \{(x, x) : 0 \leq x \leq 1\}.$$

Then

$$\int_0^1 \left[\int_0^1 f(x, y)\, d\mu_1(x) \right] d\mu_2(y) = 0$$

and

$$\int_0^1 \left[\int_0^1 f(x, y)\, d\mu_2(y) \right] d\mu_1(x) = 1$$

Note that in this case f is non-negative and the iterated integrals are unequal. Observe also, that μ_2 is not σ-finite, i.e., it's not possible to express $[0, 1]$ as

$$[0, 1] = \bigcup_{n=1}^{\infty} A_n,$$

A_n is a finite set for all n.

CHAPTER 6

Relation Between Differentiation and Integration

6.1   Differentiation

The concept of differentiation is of the main concern in this section. However, in order to define the concept of differentiation, we need to know the ideas of bounded variation and absolute continuity.

6.1.1   Bounded Variation and Absolute Continuity

1. Let f be a real-valued function defined on an interval $I = [a, b]$. For each partition $P : a = x_0 < x_2 < \ldots < x_n = b$ of $[a, b]$, let

$$V_a^b(f,p) = \sum_{i=1}^{n} |f(x_i) - f(x_{i-1})|.$$

The extended real number $V_a^b f$ given by

$$V_a^b f = \sup \left\{ V_a^b(f,p) = p \text{ is a partition of } [a,b] \right\}$$

41

is called the *total variation* of f. If $V_a^b f < \infty$, then f is called a *function of Bounded Variation* on $[a, b]$.

Example 1:

Let f be a function defined on $[0, 1]$ as follows

$$f(x) = \begin{cases} x^2 \sin\frac{1}{x}, & \text{if } x \neq 0 \\ 0, & \text{if } x = 0 \end{cases}.$$

It can be shown that f is of bounded variation as well as an *absolutely continuous* function (see below).

Example 2:

Let f be a function defined on $[0, 1]$ as follows

$$f(x) = \begin{cases} x \cos\frac{\pi}{2x}, & \text{if } x \neq 0 \\ 0, & \text{if } x = 0 \end{cases}.$$

It can be shown that f is bounded but NOT of bounded variation.

2. A function f on $[a, b]$ is said to be *absolutely continuous* if for each $\varepsilon > 0$, there exists $\delta > 0$ such that, whenever

$$\{(x_1, y_1), (x_2, y_2), \ldots, (x_n, y_n)\}$$

is a finite collection of non-overlapping sub-intervals of $[a, b]$ with

$$\sum_{i=1}^{n} |y_i - x_i| < \delta$$

then

$$\sum_{i=1}^{n} |f(y_i) - f(x_i)| < \varepsilon.$$

6.1.2 Properties of Functions of Bounded Variation

1. The equality

$$V_a^c f + V_c^b f = V_a^b f$$

holds for all $c \in [a, b]$.

2. $V_a^x f$ is a non-decreasing function of x on $[a, b]$.

3. *Characterization* of functions of bounded variation: f is a function of bounded variation if and only if it can be written as the difference of two non-decreasing functions g and h.

4. A real valued function f on $[a, b]$ of bounded variation has a finite derivative $f'(x)$ a. e.

   Remark: Since a function of bounded variation f can be expressed as a difference of two non-decreasing functions, f has thus only jump discontinuities on $[a, b]$ and they are countable in number. We say that f has a jump discontinuity of $x = x_0$ if

$$\lim_{\substack{x > x_0 \\ x \to x_0}} f(x) = F(x_0^+), \quad \lim_{\substack{x < x_0 \\ x \to x_0}} f(x) = F(x_0^-),$$

   and $f(x_0^+) \neq f\left(x_0^-\right)$.

6.1.3 Properties of Absolutely Continuous Functions

1. If f is an absolutely continuous function on $[a, b]$, then f is a function of bounded variation on $[a, b]$.

2. An absolutely continuous function on $[a, b]$ is differentiable a. e. on $[a, b]$.

3. If f is of bounded variation on $[a, b]$, then f is absolutely continuous on $[a, b]$ if and only if the total variation function $V_a^x f$ is absolutely continuous on $[a, b]$.

The following is a standard characterization of an absolutely continuous function.

4. If G is a real-valued function on $[a, b]$, then G is an absolutely continuous function on $[a, b]$ if and only if

$$G(x) = \int_a^x g(t)dt + G(a)$$

where g is an integrable function on $[a, b]$.

5. If f and g are absolutely continuous function on $[a, b]$, then

    i) $f \pm g$ is absolutely continuous.

    ii) $f \times g$ is absolutely continuous.

6.1.4 Two Theorems of Differentiation

1. Let $f : [a, b] \rightarrow \mathbf{R}$ be a non-decreasing function. Then, f is differentiable a. e.; f is a measurable function and

$$\int_a^b f'(x)dx \le f(b) - f(a).$$

2. Lebesgue Differentiation Theorem: If f is integrable on $\mathbf{R}$, its indefinite integral is differentiable with derivative $f(x)$ at almost every $x \in \mathbf{R}$, i.e.,

If $f : [a, b] \rightarrow \mathbf{R}$ is integrable and

$$F(x) = F(a) + \int_a^x f(t)dt,$$

then

$$F'(x) = f(x) \text{ a. e. in } [a, b].$$

(Recall that: differentiability is a local character.)

6.2   Introduction to the Riemann-Stieltjes Integral

In this section, the Riemann-Stieltjes integral is defined as the limit of the Riemann-Stieltjes sum.

6.2.1   Riemann-Stieltjes Sum

Let f and φ be two functions which are defined and finite on a finite interval $[a, b]$. If P $\{a = x_0 < x_1 < x_2 < \ldots < x_n = b\}$ is a partition of $[a, b]$. We arbitrarily select intermediate points

$$\left\{\xi_i\right\}_{i=1}^n$$

satisfying

$$x_{i-1} \le \xi_i \le x_i,$$

and write

$$R_P = \sum_{i=1}^n f(\xi_i)\big(\varphi(x_i) - \varphi(x_{i-1})\big).$$

R_P is called a *Riemann-Stieltjes sum* for P.

6.2.2   Riemann-Stieltjes Integral

Let f, φ, P, and R_P be as in 1 above. If

$$I = \lim_{|P| \to 0} R_Q, \; |P| = \max\left\{(x_1 - x_0), \; \ldots, (x_n - x_{n-1})\right\}$$

exists and is finite, that is, given $\varepsilon > 0$, there is a $\delta > 0$ such that $| I - R_P | < \varepsilon$ for any P satisfying $| P | < \delta$, then I is called the *Riemann-Stieltjes integral* of f with respect to φ on $[a, b]$, and denoted by

$$I = \int_a^b f(x) \, d\varphi(x) = \int_a^b f d\varphi.$$

6.2.3 Remarks on Riemann-Stieltjes Integral

1. If $\varphi(x) = x$,

$$\int_a^b f \, d\varphi$$

is just the Riemann integral

$$\int_a^b f \, dx.$$

2. If f is continuous on $[a, b]$ and φ is continuously differentiable on $[a, b]$, then

$$\int_a^b f \, d\varphi = \int_a^b f \varphi' dx.$$

3. More generally:

    a. If f is continuous on $[a, b]$ and φ is absolutely continuous on $[a, b]$, then

    $$\int_a^b f \, d\varphi = \int_a^b f \varphi' dx$$

    b. If both f and φ are absolutely continuous on $[a, b]$, then

    $$\int_a^b f \, \varphi' dx = \varphi(b) f(b) - \varphi(a) f(a) - \int_a^b \varphi' f dx.$$

6.3 Introduction to the Lebesgue-Stieltjes Integral

The Lebesgue-Stieltjes integral is defined via the Lebesgue-Stieltjes outer measure. However, in spite of this difference from the Riemann-Stieltjes integral, a simple relationship exists between these two types of integrals.

6.3.1 Lebesgue-Stieltjes Measure

Let f be a fixed finite and monotone increasing function on $(-\infty, \infty)$. For each half-open finite interval of the form $(a, b]$. Let

$$\lambda(a, b] = \lambda_f((a, b]) = f(b) - f(a).$$

Observe that $\lambda \geq 0$ since f is increasing. Let A be a non-empty subset of **R**, and

$$\Lambda^* (A) = \Lambda^*_f(A) = \inf \Sigma \lambda [a_k, b_k]$$

where the inf is taken over countable collections $\{[a_k, b_k]\}$ such that

$$A \subset \bigcup_{\mathbf{R}} (a_k, b_k)$$

Further, define $\Lambda^* (\phi) = 0$.

Λ^*_f is called the *Lebesgue-Stieltjes outer measure* corresponding to f, and its restriction to those sets which are Λ^*_f-measurable is called the *Lebesgue-Stieltjes measure* corresponding to f, and denoted by Λ_f or simply Λ.

Example 1: The Lebesgue-Stieltjes outer measure Λ^*_f corresponding to $f(x) = x$ coincides with ordinary Lebesgue outer measure. Moreover, a set A is Λ^*_f-measurable if and only if it is Lebesgue measurable[1].

Example 2: If μ is a finite Borel measure (see section 3.3.7) on **R**, define

$$f_\mu(x) = \mu([-\infty, x]), -\infty < x < \infty$$

observe that f_μ is monotone increasing and that $\mu([a, b]) = f_\mu(b) - f_\mu(a)$. It follows that the Lebesgue-Stieltjes outer measure induced by f_μ agree with μ as a Borel measure.

6.3.2 Lebesgue-Stieltjes Integral

Let g be a Borel measurable function defined on $\mathbf{R}$, and let Λ_f be a Lebesgue-Stieltjes measure. Then the integral

$$\int g d\Lambda_f$$

is called the *Lebesgue-Stieltjes integral* of g with respect to Λ_f.

6.3.3 Riemann-Stieltjes vs. Lebesgue-Stieltjes Integral

Let f be an increasing function which is right continuous on $[a, b]$ and let g be a bounded Borel measurable function on $[a, b]$. If the Riemann-Stieltjes integral

$$\int_a^b g \, df$$

exists, then

$$\int_{(a,b]} g d\Lambda_f = \int_a^b g \, df.$$

[1] *This follows from Carathéodory's theorem. Refer to section 3.3.5.*

CHAPTER 7

The L^p – Spaces

7.1 Definition of the L^p – Spaces

Let p be a positive real number, and let f be a measurable function defined on $[0, 1]$ such that

$$\int_0^1 |f(x)|^p \, dx < \infty.$$

Then we say that f belongs to the class $L^p[0, 1]$ or in short L^p. The *norm* $\|f\|_p$ is defined by

$$\|f\|_P = \left(\int_0^1 |f(x)|^p \, dx \right)^{\frac{1}{p}}$$

Remarks on L^p:

1. One can study the L^p spaces in a more general and abstract framework, where the interval $[0, 1]$ is replaced by some abstract set E, the measurable subsets of $[0, 1]$ by a certain family of subset of E, and the Lebesgue integration by a more general process. Most of the theorems and proofs for the case of $[0, 1]$ follow in exactly the same way in the abstract case.

2. If two functions in L^p coincide except on a set of measure zero, we shall consider them to represent the same element of L^p. In other words, if we write $f \sim g$, *iff* $f = g$ a. e., then clearly $\sim$ is an equivalence relation and the elements of L^p are actually equivalence classes of functions, the functions in any one class differing from one another only on sets of measure zero. We shall refer to a function in L^p instead of the equivalence class containing f. Henceforth we shall be concerned with L^p for values of $P \geq 1$ exclusively.

7.2   The L^∞ – Spaces

A measurable function defined on $[0, 1]$ is called *essentially bounded* if and only if there is some $M \geq 0$ such that the set $\{x : f(x \geq M\}$ has measure zero. In other words, f is essentially bounded if and only if the inequality $|f(x)| \leq M$ holds almost everywhere on $[0, 1]$ for some M. The *norm* $\|f\|_\infty$ of an essentially bounded function is defined by

$$\|f\|_\infty = \inf\Big\{M : m\big\{x : |f(x)| \geq M\big\} = 0\Big\}.$$

The number $\|f\|_\infty$ is also called the *essential supremum* of f.

Remarks on $L\infty$:

1. A direct consequence of the definition of $\|f\|_\infty$ is that for every $\in > 0$ the set

$$\Big\{x : |f(x)| > \|f\|_\infty - \in\Big\}$$

has positive measure.

2. $L\infty$ denotes the class of all measurable and essentially bounded functions on $[0, 1]$. Here also, as in the case of the L^p spaces, we identify two functions in $L\infty$ if they agree almost everywhere.

7.3 Properties of the L^p – Spaces ($p \geq 1$ and $p = \infty$)

1. If $a \in \mathbf{R}$, then $af \in L^p$ whenever $f \in L^p$.

2. If $f, g \in L^p$, then $f + g \in L^p$. Because the inequality

    $$|f + g|^p \leq 2^p(|f| + |g|^p)$$

    is true.

3. Let $1 \leq p \leq q$. Then, if $f \in L^q$, we have $f \in L^p$. But L^p is not necessarily in L^q. For example, $p = 1$ and $q = 2$.

    $$\text{If } f(x) = \frac{1}{\sqrt{x}},$$

    then $f \in L^1$ ([0, 1]), but $f \notin L^2$ ([0, 1]).

4. (L^p, d), $d(f, g) = \|f - g\|_p$ is a metric space.

7.4 The Holder's Inequality

Let p and q be real numbers such that $1 \leq p \leq +\infty$ and

$$\left(\frac{1}{p}\right) + \left(\frac{1}{q}\right) = 1.$$

Let $f \in L^p$ and $g \in L^q$. Then

$$fg \in L' \text{ and}$$

$$\int_0^1 |fg| \leq \|f\|_p \|g\|_q.$$

Equality holds if and only if there exists constants a and b not both zero such that $a|f(x)|^p = b|g(x)|^q$ a. e.

7.5   The Minkowski's Inequality

If f and g belong to L^p, $p \geq 1$, then

$$\|f + g\|_p \leq \|f\|_p + \|g\|_p.$$

7.6   Convergence in the Mean of Order p $(1 < \ \leq p < +\infty)$

1. Let $\{f_n\}$ be a sequence of functions in L^p, $(1 \leq p < \infty)$, and let $f \in L^p$. We say that $\{f_n\}$ converges to f in the mean of order p (or that $\{f_n\}$ converges to f in L^p or that $\{f_n\}$ is L^p–convergent to f) if and only if

$$\lim_{n \to \infty} \|f_n - f\|_p = 0.$$

2. A sequence $\{f_n\}$ in L^p is said to be a cauchy sequence in L^p if and only if for every $\in > 0$ there exists an integer N such that $n \geq N$, $m \geq N$ implies

$$\|f_n - f_m\|_p < \in$$

Remark: The notion of convergence in L^p is different from the notion of pointwise convergence.

Example: Let f_n be defined on $[0, 1]$ as follows

$$f_n(x) = \begin{cases} n, & \text{if } x \in \left[0, \dfrac{1}{n}\right] \\ 0, & \text{if } x \in \left[\dfrac{1}{n}, 1\right] \end{cases}.$$

Then, f_n converges pointwise to the function identically 0 on $[0, 1]$. But

$$\|f_n - 0\|_1 = \int_0^1 f_n(x)\, dx = \int_0^{\frac{1}{n}} n\, dx = 1 \neq 0$$

So, f_n does not converge to 0 in L^1.

7.7 Completeness (Theorem)

The space L^p is a complete metric space for $1 \le p < \infty$.

7.8 Bounded Linear Functionals on L^p

1. Let F be a real-valued (or complex-valued) function defined on L^p such that

    $$F(af + bg) = aF(f) + bF(g)$$

    for all f and g in L^p and for all real numbers a and b. Then F is called a *Linear functional* on L^p.

2. A Linear functional F on L^p is said to be *bounded* if and only if there is a positive constant M such that

    $$|F(f)| \le M\|f\|_p \text{ for all } f \in L^p.$$

3. The norm $\|F\|$ of a bounded Linear functional F is defined to be the infimum of the set of all the numbers M satisfying the inequality in statement 2 above, i.e.,

    $$\|F\| = \inf\left\{ M : |F(f)| \le M\|f\|_p,\, f \in L^p \right\}.$$

    Observe that

    $$|F(f)| \le \|F\|\, \|f\|_p \text{ for all } f \in L^p.$$

    The following theorem gives an important class of linear functionals on L^p.

7.8.1 Two Theorems of Bounded Linear Functional on L^p

1. Theorem

   Let p and q be extended real numbers such that $1 \leq p \leq +\infty$ and

   $$\left(\frac{1}{p}\right) + \left(\frac{1}{q}\right) = 1.$$

   If $g \in L^q$ and F_g is defined by

   $$\left|F_g(f)\right| \leq \|F\| \, \|f\|_p \text{ for all } f \in L^p.$$

   then F_g is a bounded linear functional on L^p. We also have

   $$\left\|F_g\right\| = \|g\|_q.$$

   Remark: When $1 \leq p < +\infty$, the converse of the above theorem is the well-known Riesz theorem.

2. Theorem (F. Riesz):

   Let p and q be real numbers such that

   $$1 \leq p < +\infty, \, 1 \leq q \leq \infty, \text{ and}$$

   $$\left(\frac{1}{p}\right) + \left(\frac{1}{q}\right) = 1.$$

   Let F be a bounded linear functional on L^p. Then there is a unique $g \in L^q$ such that

   $$F_g(f) = \int_0^1 f(x) \, g(x) \, dx, \text{ for all } f \in L^p.$$

We have also

$$\|F\| = \|g\|_q.$$

Remark: If $p = \infty$, not every linear functional can be expressed as

$$\int_0^1 fg.$$

Example: For $f \in C([0,1])$, let $G(f) = f(1)$. Then G is a bounded linear functional on $C([0, 1])$ and extends to an F, a bounded linear functional on $L^\infty ([0, 1])$. This follows from the Hahn-Banach Theorem (discussed in chapter 9).

For $n = 1, 2, \ldots$, let

$$f_n(x) = \begin{cases} 0, \text{ if } 0 \le x \le 1 - \dfrac{1}{n} \\ nx - n + 1, \text{ if } 1 - \dfrac{1}{n} < x \le 1 \end{cases}$$

Then, $F(f_n) = G(f_n) = 1$, but for any $g \in L^1([0, 1])$,

$$\lim_{n \to \infty} \int_0^1 f_n g\,dx = 0$$

so that $F \ne F_g$ for any $g \in L^1 ([0, 1])$.

7.9   The Dual of L^p

Definition: Let us denote by $(L^p)^*$ the set of all bounded linear functionals on L^p where $1 \le p < \infty$. $(L^p)^*$ is called the *dual* or the *conjugate space* of L^p.

Properties:

1. $(L^p)^*$ is a complete normed linear space. A complete normed linear space is called a *banach space*. So, both L^p and $(L^p)^*$ are banach spaces.

2. For $1 \leq p < \infty$, by theorems in section 7.6, $(L^p)^* = L^q$, where

$$\left(\frac{1}{p}\right) + \left(\frac{1}{q}\right) = 1.$$

But, we have seen, by means of the example in section 7.8.1 above, that $(L^\infty)^* \neq L^1$, though $(L^1)^* = L^\infty$.

CHAPTER 8

Banach Spaces

8.1 Definition of Banach Spaces

1. A *normed linear space* is a linear space **N** in which to each vector x there corresponds a real number, denoted by $\|x\|$ and called the *norm* of x, in such a manner that

    i) $\|x\| \geq 0$, and $\|x\| = 0$ if $x = 0$;

    ii) $\|x + y\| \leq \|x\| + \|y\|$;

    iii) $\|dx\| = |d|\,\|x\|$, for all $d \in \mathbf{R}$ or $d \in \mathbf{C}$.

    Remark: A normed linear space **N** is a metric space with respect to the metric d defined by

    $$d(x,y) = \|x - y\|.$$

2. A *Banach space* is a complete normed linear space (complete as a metric space with respect to the metric $d(x,y) = \|x - y\|$).

8.2 Examples of Banach Spaces

1. The linear spaces $\mathbf{R}^n$ and $\mathbf{C}^n$ of n-tuples $x = (x_1, x_2, \ldots, x_n)$ of real and comlex numbers, respectively, can be made into normed linear spaces an infinite variety of ways. In particular, if the norm is defined by

    $$\|X\| = \left(\sum_{i=1}^{n} |x_i|^2 \right)^{\frac{1}{2}},$$

    then we get the familiar n-dimensional *Euclidean* and *Unitary* spaces, respectively.

2. Let p be a real number such that $1 \le p < \infty$, and we denote by ℓ_p the space of all sequences

    $$X = (x_1, x_2, \ldots, x_n, \ldots)$$

    of scalars such that

    $$\sum_{n=1}^{\infty} |x_n|^p < \infty,$$

    with the norm defined by

    $$\|X\|_p = \left(\sum_{n=1}^{\infty} |x_n|^p \right)^{\frac{1}{p}}.$$

3. All the L^p spaces, $1 \le p \le \infty$, discussed in Chapter 7 with norm

    $$\|f\|_p = \left(\int |f(x)|^p \, dx \right)^{\frac{1}{p}}$$

are examples of Banach spaces. Also, L^∞ is a Banach space with norm defined by

$$\|f\|_\infty.$$

4. The space $C(x)$ of all bounded continuous real or complex valued functions defined on a topological space X (say X is a metric of a normed linear space), with the norm given by

$$\|f\| = \sup_{x \in X} |f(x)|$$

is a Banach space. This norm is sometimes called a *uniform norm*, because the statement that f_n converges to f with respect to this norm means that f_n converges to f uniformly on X.

Remark: In the spaces discussed in Example 2, i.e., ℓ_p spaces, the following inequalities are true:

i) Hölder's Inequality:

$$\sum_{i=1}^{\infty} |x_i y_i| \le \|x\|_p \|y\|_q, \quad \left(\frac{1}{p}\right) + \left(\frac{1}{q}\right) = 1.$$

ii) Minkowski's Inequality:

$$\|X + Y\|_p \le \|X\|_p + \|Y\|_p.$$

8.3 Separability

Definition: A topological space (metric space, normed linear space) is called *separable* if it has a countable dense subset.

Theorem: If the conjugate (dual) space N^* of a normed linear space N is separable, then N is separable.

Remark: The converse of the above theorem is not necessarily true, i.e., **N** could be separable but **N*** may not be, as the example below shows:

Example:

$$\text{Let } \mathbf{N} = \left\{ (x_1, x_2, ..., x_n, ...) : \|X\|_1 < \infty, \text{ and } x_i \in \mathbf{R} \right\},$$

$$\|X\|_1 = \sum_{i=1}^{\infty} |x_i|.$$

This space is usually denoted by ℓ_1.

N is separable, since the subset of all elements $(x_1, x_2, ..., x_n, ...)$ of **N**, where each x_i is a rational number, is countable and dense in **N**.

However, $\left(\ell_1 \right)^* = \ell_\infty$, i.e., the normed linear space whose underlying set is the collection of all bounded sequences of the form $X = (x_1, x_2, ..., x_n, ...)$ and $\|X\|_\infty$ is defined by

$$\|X\|_\infty = \sup |x_i|.$$

But ℓ_∞ contains the uncountable set.

$S = \{(x_1, x_2, ..., x_n, ...) : x_i \text{ is 1 for a finite number of } i\text{'s and 0 otherwise}\}$.

Moreover, for two distinct x_i, x_i in S

$$\left\| x_i - x_j \right\|_\infty = 1.$$

So, no countable subset of ℓ_∞ can be dense in ℓ_∞, since if we take any countable subset A of ℓ_∞ and take an open ball of radius $\frac{1}{2}$ centered at each element of A, then each ball contains at most one element of S.

8.4   The Unit Sphere

Many important properties of a Banach space are closely linked to the shape of its closed *unit sphere*, that is, the set

$$S = \left\{ x : \|X\| \leq 1 \right\}.$$

8.4.1   Properties of the Unit Sphere

The closed unit sphere is convex, in the sense that, if X and Y are any two vectors in S, then the vector $Z = aX + bY$ is also in S, where a and b are non-negative real numbers such that $a + b = 1$, for

$$\|Z\| = \|aX + bY\| \leq a\|X\| + b\|Y\| \leq a + b = 1.$$

In this connection, it is illuminating to consider the shape of S for certain simple examples:

Illustrations: Let the underlying linear space be $\mathbf{R}^2$. The figure below shows the closed unit ball for four different norms defined on $\mathbf{R}^2$:

$\|n\|_2 = 1$

$\|x\|_2 = \left(x_1^2 + x_2^2 \right)^{\frac{1}{2}}$

$\|n\|_\infty = 1$

$= \max\left\{ |x_1|, |x_2| \right\}$

$\|x\|y_2 = 1$

$\|x\|y_2 = \left(|x_1|^{\frac{1}{2}} + |x_2|^{\frac{1}{2}} \right)^2$

$\|x\|_1 = 1, \|x\|_1 = |x_1| + |x_2|$

The figure above illustrates the four closed unit spheres corresponding to the four different norms. When the norm is one of $\|\cdot\|_1$, $\|\cdot\|_2$, or $\|\cdot\|_\infty$, the corresponding closed unit sphere is convex. However, if we define $\|X\|_{\frac{1}{2}}$ by $\|X\|_{\frac{1}{2}} = \left(|x_1|^{\frac{1}{2}} + |x_2|^{\frac{1}{2}} \right)^2$, then as the dotted part of the figure above shows

$$\left\{ x : \|X\|_{\frac{1}{2}} \leq 1 \right\}$$

is not convex. So, for $p < 1$, the formula

$$\|X\|_p = \left(|x_1|^p + |x_2|^p \right)^{\frac{1}{p}}$$

does not yield a norm.

CHAPTER 9

Linear Operators

9.1   Definition of Linear Operators

Let L and L' be linear spaces with the same system of scalars (say the system of scalars is $\mathbf{R}$ or $\mathbf{C}$). A mapping T of L into L' is called a *linear transformation* if

$$T(x + y) = T(x) + T(y) \text{ and } T(ax) = aT(x)$$

or equivalent, if

$$T(ax + by) = aT(x) + \text{b}T(y)$$

for all vectors x, y in L and scalars a, b.

Remark: If $L' = \mathbf{R}$ or $\mathbf{C}$, then T is called a linear functional.

Examples:

1. We consider the linear space $\mathbf{R}^3$, and each linear transformation mentioned is a mapping of $\mathbf{R}^3$ into itself:

    i) $T_1[(x_1, x_2, x_3)] = (cx_1, cx_2, cx_3)$ where c is a fixed number.

    ii) $T_2[(x_1, x_2, x_3)] = (x_1, x_2, 0)$. T_2 projects $\mathbf{R}_3$ onto the $X - Y$ plane.

2. The mapping $I : ([0, 1]) \rightarrow \mathbf{R}$, defined by

$$I(f) = \int_0^1 f(x)\,dx$$

is easily seen to be a linear transformation of L' ([0, 1]) into the real linear space $\mathbf{R}$ of real numbers.

3. Consider the linear space $\mathcal{P}$ of all polynomials $P(x)$ with real coefficients, defined on [0, 1]. The mapping D defined by

$$D(p) = \frac{dP}{dx}$$

is clearly a linear transformation of $\mathcal{P}$ into itself.

9.2   Properties of Linear Operators

Let $\mathbf{N}$ and $\mathbf{N}'$ be normal linear spaces and T a linear transformation of $\mathbf{N}$ into $\mathbf{N}'$. Then, the following conditions on T are all equivalent to one another.

1. T is continuous;

2. T is continuous at one origin in the sense that $x_n \rightarrow 0$ implies $T(x_n) \rightarrow 0$;

3. There exists a real number $K \geq 0$ with the property that

$$\|T(x)\| \leq K\|x\| \text{ for every } x \in \mathbf{N};$$

4. If $s = \left\{x : \|x\| \leq 1\right\}$ is the closed unit sphere in $\mathbf{N}$, then its image $T(S)$ is a bounded set in $\mathbf{N}'$.

9.3   Boundedness of Linear Transformations

If the linear transformation T in the above theorem satisfies statement 3, so that there exists a real number $K \geq 0$ such that

$$\|T(x)\| \leq K\|x\|$$

for every x, then K is called a *bound* for T, and such a T is often referred to as a *bounded linear transformation*.

Note, according to section 9.2, T is bounded if and only if it is continuous, so these two terms can be used interchangeably.

9.3.1 Norm of a Continuous Operator

If T is continuous, its norm is defined by

1. $\|T\| = \sup\left\{\|T(x)\| : \|\mathbf{x}\| \leq 1\right\}.$

   when $\mathbf{N} \neq \{0\}$. This formula has a useful equivalent form

2. $\|T\| = \sup\left\{\|T(x)\| : \|\mathbf{x}\| = 1\right\}.$

   In addition, we can also have

3. $\|T\| = \inf\left\{K : K \geq 0 \text{ and } \|T(x)\| \leq K\|x\|, \text{ for all } x\right\}.$

   Then, we can easily see that

$$\|T(x)\| \leq \|T\|\,\|x\|.$$

9.3.2 A Theorem

If $\mathbf{N}$ and $\mathbf{N}'$ are normed linear spaces, then the set $\mathcal{B}\,(\mathbf{N}, \mathbf{N}')$ of all continuous linear transformations of $\mathbf{N}$ into $\mathbf{N}'$ is itself a normed linear space with respect to the pointwise linear operations and the norm defined by statement 1 above. Further, if $\mathbf{N}'$ is a Banach space, then $\mathcal{B}\,(\mathbf{N}, \mathbf{N}')$ is also a Banach space.

9.4 Two Examples

1. Let $\mathbf{N}$ be a finite-dimensional normed linear space with dimension $n > 0$, and let $\{e_1, e_2, \ldots, e_n\}$ be a basis for $\mathbf{N}$. Each vector x in $\mathbf{N}$ can be written uniquely in the form

$$x = \alpha_1 \, e_1 + \alpha_2 \, e_2 + \ldots + \alpha_n \, e_n,$$

for some $\alpha_i \in \mathbf{R}$ (or $\mathbf{C}$), $i = 1, 2, \ldots, n,$ not all zero. If T is the linear transformation of $\mathbf{N}$ onto

$$\mathbf{N}' = \left\{ (a_1, \, a_2, \, \ldots \, a_n) : a_1, \, a_2, \, \ldots \, a_n \in \mathbf{R} \right\}$$

and

$$\left\| (a_1, \, \ldots, \, a_n) \right\| = |a_1| + \ldots + |a_n|,$$

defined by

$$T(x) = (\alpha_1, \, \alpha_2, \, \ldots, \, \alpha_n).$$

then T is one-to-one into $\mathbf{N}'$ and continuous. In fact, T^{-1} is also continuous.

2. The bounded linear functional on L^p space, as defined in section 7.8 of Chapter 7.

9.5   The Hahn-Banach Theorem

If $\mathbf{N}$ is a normed linear space over $\mathbf{R}$, the set of all continuous linear transformations of $\mathbf{N}$ into $\mathbf{R}$, i.e., $\mathcal{B}$ ($\mathbf{N}$, $\mathbf{R}$) is denoted by $\mathbf{N}^*$ and is called the *conjugate space* of $\mathbf{N}$. The elements of $\mathbf{N}^*$ are called *continuous linear functionals*. If $\mathbf{N}$ is over $\mathbf{C}$, then $\mathbf{N}^*$ denotes $\mathcal{B}$ ($\mathbf{N}$, $\mathbf{C}$). For example, for $1 \leq P < \infty$, we have seen in Chapter 7, that

$$\left(L^p \right)^* = L^q, \; \left(\frac{1}{p} \right) + \left(\frac{1}{q} \right) = 1.$$

The Hahn-Banach Theorem:

Let M be a linear subspace of a normed linear space $\mathbf{N}$, and let f be a functional defined on M. Then, f can be extended to a functional to defined on the whole space $\mathbf{N}$ such that

$$\|f_0\| = \|f\|.$$

Note that the Hahn-Banach theorem guarantees that any Banach space (or normed linear space) has a rich supply of functionals. Most applications of the Hahn-Banach theorem depend on the following consequences of the theorem.

1. If **N** is a normed linear space and X_0 is a non-zero vector in **N**, then there exists a functional f_0 in **N*** such that

    $$f_0(X_0) = \|X_0\| \text{ and } \|f_0\| = 1.$$

2. If M is a closed linear subspace of a normed linear space **N** and X_0 is a vector not in M, then there exists a functional f_0 in **N*** such that $f_0(M) = 0$ and $f_0(X_0) \neq 0$.

    Both 1 and 2 can be regarded as theorems themselves. Furthermore, as a consequence of statement 2, we have

3. Let M and **N** be as given in statement 2. If d is the distance from X_0 to M, then there exists a functional f_0 in **N*** such that

    $$f_0(M) = 0, \ f_0(X_0) = 1, \text{ and } \|f_0\| = \frac{1}{d}.$$

(Recall that distance of X_0 to M is $\inf\{\|X_0 + m\| = m \in M\}$.)

CHAPTER 10

Hilbert Space

10.1 Hilbert Space

A *Hilbert space H* is a complex Banach space whose norm arises from an *inner product*, $< x, y >$, which defines a complex function of vectors x and y and has the following properties:

i) $< ax + by, z > = a < x, z > + b < y, z >$;

ii) $< x, y > = \overline{< y, x >}$, where the bar sign means complex conjugate;

iii) $< x, x > = \|x\|^2$,

for all x, y, z in H and complex numbers a and b.

10.1.1 Some Examples of Hilbert Space

1. $\mathbf{C}^n$ with norm defined by

$$\|(x_1, x_2, \ldots, x_n)\|^2 = |x_1|^2 + |x_2|^2 + \ldots + |x_n|^2$$

where

$$|x_i|^2 = x_i \overline{x}_i, \ i = 1, \ 2, \ ..., \ n.$$

And, the inner product of *two vectors* x and y defined by

$$< x, y > = \sum_{i=1}^{n} x_i \overline{y}_i.$$

2. The space ℓ_2 with the inner product of two vectors $x = (x_1, x_2, ..., x_n, ...)$ and $(y = y_1, y_2, ..., y_n, ...)$ defined by

$$< x, y > = \sum_{i=1}^{\infty} x_i \overline{y}_i$$

Recall that

$$\ell^2 = \left\{ (x_1, \ x_2, \ ..., \ x_n, \ ...) : x_i \in C \text{ and } \sum_{i=1}^{\infty} |x_i|^2 < \infty \right\}.$$

3. The space L^2 with a measure space X, measure m and inner product of two functions f and g is defined by

$$< f, g > = \int f(x) \ \overline{g(x)} \ dm(x).$$

10.1.2 Properties of Hilbert Space

1. Schwarz's Inequality (Theorem):

If x and y are any two vectors in a Hilbert space, then

$$|< x, y >| \le \|x\| \ \|y\|.$$

Proof: If $y = 0$, the result is trivial. If $y \ne 0$, the inequality is equivalent to

$$\left| < x, \frac{y}{\|y\|} > \right| \le \|x\|.$$

We may, therefore, confine our attention to proving: if $\|y\| = 1$, then we have $|< x, y >| \le \|x\|$. But this is a direct consequence of the fact that

$$0 \le \|x - < x, y > y\|^2 = \|x\|^2 - |< x, y >|^2.$$

The following figure shows the geometrical interpretation of the different quantities in this proof.

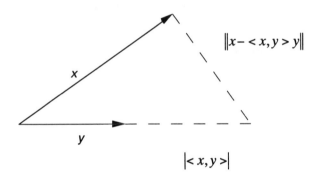

2. Parallelogram Law:

This law states that for two vectors x, y in a Hilbert space

$$\|x + y\|^2 + \|x - y\|^2 = 2\|x\|^2 + 2\|y\|^2.$$

As a consequence of the parallelogram law, we have the following important theorem:

3. Theorem:

A closed convex subset C of a Hilbert space H contains a unique vector of the smallest norm.

10.2 An Important Criterion (Theorem)

If B is a complex Banach space *whose norm obeys the parallelogram law,* and if an inner product is defined on B by

$$<x,y> = \frac{1}{4}\left(\|x+y\|^2 - \|x-y\|^2 + i\|x+iy\|^2 - i\|x-iy\|^2\right),$$

then B is a Hilbert space.

10.3 Orthogonal Complements

The following five theorems discuss the concepts of projection and orthogonality in the Hilbert space.

1. Theorem:
   Let M be a closed linear subspace of a Hilbert space H, let x be a vector not in M, and let d be the distance from x to M, i.e.,

   $$d = \inf\left\{\|x+m\| : m \in M\right\}.$$

   Then, there exists a unique vector y_0 in M such that

   $$\|x - y_0\| = d$$

   (see figure below).

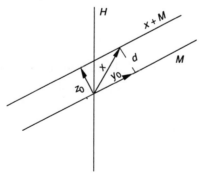

y_0 is called the "projection of x on M."

2. Theorem:
   If M is a proper closed linear subspace of a Hilbert space H, then there exists a non-zero vector z_0 in H such that $z_0 \perp M$, i.e., $<z_0, y> = 0$ for all $y \in M$ (see figure above).

3. Theorem:
   If M and $\mathbf{N}$ are closed linear subspaces of a Hilbert space H such that $M \perp N$, then the linear subspace $M + N$ is also closed.

   The following is the main theorem on orthogonality and follows from the above three theorems.

4. Theorem:
   If M is a closed linear subspace of a Hilbert space H, then $H = M \oplus M^\perp$ (i.e., every element v of H can be uniquely written as $v = x + y$, with $x \in M$ and $y \in M^\perp$), where

   $$M^\perp = \{y : y \perp M, \text{ i.e., } <y, m> = 0 \text{ for all } m \in M\}.$$

   This theorem guarantees the existance of projections in a Hilbert space.

5. Theorem:
   If M is an arbitrary closed linear subspace of a Hilbert space H, then the last theorem shows that there exists a projection defined on H whose range is M and whose null space is $M^\perp$.

10.4 Orthonormal Sets

An *orthonormal set* in a Hilbert space H is a non-empty subset of H which consists of mutually orthogonal unit vectors; that is, it is a non-empty subset $\{e_i\}$ of H with the following properties:

   i) $i \neq j$ implies $e_i \perp e_j$;

   ii) $\|e_i\| = 1$ for every i.

10.4.1 Two Examples

1. The subset $\{e_1, e_2, ..., e_n\}$ of ℓ_n^2, where e_i, $i = 1, 2, ... n$, is the n-tuple with 1 in the ith place and 0's elsewhere, is evidently an orthonormal set in this space. Recall that

$$\ell_n^2 = \left\{(x_1, x_2, ... \, x_n) : x_i \in \mathbf{C}, \|x\|^2 = \sum_{i=1}^{n} |x_i|^2\right\}.$$

2. If $H = \ell^2$ and e_n is the sequence with 1 in the nth place and 0's elsewhere, then $\{e_1, e_2, ..., e_n, ...\}$ is an orthonormal set in ℓ^2.

10.4.2 Important Theorems on Orthonormal Sets

1. Theorem:
   Let $\{e_1, e_2, ..., e_n\}$ be a finite orthonormal set in a Hilbert space H. If x is any vector in H, then

$$\sum_{i=1}^{n} |<x, e_i>|^2 \leq \|x\|^2;$$

   moreover,

$$\left(x - \sum_{i=1}^{n} <x, e_i> e_i\right) \perp e_j \text{ for every } j.$$

2. Theorem:
   If $\{e_i\}$ is an orthonormal set in a Hilbert space H, and if x is any vector in H, then the set

$$S = \{e_i : <x, e_i> \neq 0\}$$

   is either empty or countable.

These two theorems result in the following well-known Bessel's inequality.

10.5 Bessel's Inequality

If $\{e_i\}$ is an orthonormal set in a Hilbert space, then

$$\sum |<x,\, e_i>|^2 \leq \|x\|^2$$

for *every vector x* in *H*.

Now, we can also generalize the second part of statement 1 in section 10.4.2 as:

10.5.1 Theorem (Arbitrary Vector)

If $\{e_i\}$ is an orthonormal set in a Hilbert space *H*, and. if *x* is an *arbitrary* vector in *H*, then

$$(x - \Sigma <x,\, e_i>) > e_i \perp e_j$$

for all *j*.

10.5.2 Complete Orthonormal Set

An orthonormal set $\{e_i\}$ in *H* is said to be *complete* if it is impossible to adjoin a vector *e* to $\{e_i\}$ in such a way that $\{e\} \cup \{e_i\}$ is an orthonormal set which properly contains $\{e_i\}$.

Theorem: Every non-zero Hilbert space contains a complete orthonormal set.

10.5.3 Theorem

Let *H* be a Hilbert space, and let $\{e_i\}$ be an orthonormal set in *H*. Then, the following conditions are equivalent to each other:

1. $\{e_i\}$ is complete.

2. $x \perp \{e_i\} \Rightarrow x = 0.$

3. If x is an arbitrary vector in H, then $x = \Sigma < x, \{e_i\} > \{e_i\}$.

4. If x is an arbitrary vector in H, then

$$\|x\|^2 = \sum |< x,\ e_i >|^2.$$

Remark: The numbers $<x,\ e_i>$, in statement 3 above, are called the *Fourier coefficients* of x, the expression $x = \Sigma < x, e_i > e_i$ is called the *Fourier Expansion* of x, and the equation

$$\|x\|^2 = \sum |< x,\ e_i >|^2$$

is called the *Parseval's Equation*, with respect to the corresponding orthonormal set $\{e_i\}$.

10.5.4 Fourier Expansion

Consider the Hilbert space L^2 associated with the Lebesgue measure on $[0, 2\pi]$.

$$L^2 = \left\{ f \text{ measurable on } [0,\ 2\pi] : \int_0^{2\pi} |f(x)|^2\ dx < \infty \right\}$$

Its norm is defined by

$$\|f\| = \left(\int_0^{2\pi} |f(x)|^2\ dx \right)^{\frac{1}{2}},$$

and inner product by

$$< f,\ g >= \int_0^{2\pi} f(x)\ \overline{g(x)}\ dx.$$

It follows that

$$\left\{ \frac{e^{inx}}{\sqrt{2\pi}}, \ n = 0, \ \pm 1, \ \pm 2, \ \pm 3, \ ... \right\}$$

form an orthonormal set in L^2, in fact, a complete one.

Moreover, if f is any function in L^2, the numbers

$$a_n = \frac{1}{\sqrt{2\pi}} \int_0^{2\pi} f(x) \, e^{-inx} dx, \ n = 0, \ \pm 1, \ \pm 2, \ ...$$

are its Fourier coefficients, and Bessel's Inequality takes the form

$$\sum_{n=-\infty}^{\infty} |a_n|^2 \le \int_0^{2\pi} |f(x)|^2 dx.$$

In addition, by the theorem in section 10.5.3 above, each f in L^2 has a Fourier expansion:

$$f(x) = \frac{1}{\sqrt{2\pi}} \sum_{n=-\infty}^{\infty} a_n \, e^{inx},$$

in the sense that

$$\|f_n - f\|_2 \to 0, \text{ where } f_n(x) = \frac{1}{\sqrt{2\pi}} \sum_{k=-n}^{n} a_k \, e^{ikx}.$$

10.5.5 Riesz-Fischer Theorem

If a_n, $n = 0, \pm 1, \pm 2, ...$ are given complex numbers for which

$$\sum_{n=-\infty}^{\infty} |a_n|^2$$

converges, i.e.,

$$\sum_{n=-\infty}^{\infty} |a_n|^2 < \infty,$$

then there exists a function f in L^2 whose Fourier coefficients are the a_n's.

10.5.6 Riemann-Lebesgue Lemma

Let $f \in L^2 [0, 2\pi]$ and

$$a_n = \frac{1}{\sqrt{2\pi}} \int_0^{2\pi} f(x) e^{-inx} dx,$$

then

$$\lim_{n \to \infty} a_n = 0.$$

In addition, if a_n is above and $a_n = 0$ for $n = 0, \pm 1, \pm 2, \ldots$, then $f = 0$, a. e. on $[0, 2\pi]$.

10.6 A Banach Space Which Is Not A Hilbert Space

Let $\mathbf{R}^2$ be equipped with the norm

$$\|(x_1, x_2)\| = |x_1| + |x_2|.$$

Obviously this makes $\mathbf{R}^2$ a Banach space. But, if we take the convex subset C, as in the figure below, there is no unique element of **C** with the minimum norm and by the theorem in section 10.1.2 this norm will not make $\mathbf{R}^2$ a Hilbert space.

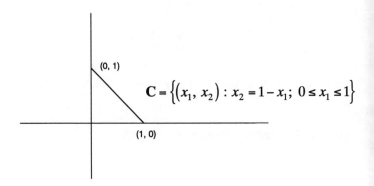

$$C = \left\{ (x_1,\, x_2) : x_2 = 1 - x_1;\ 0 \le x_1 \le 1 \right\}$$

Observe that on **C** every point (x_1, x_2) has norm

$$\left\| (x_1,\, x_2) \right\| = |x_1| + |x_2| = x_1 + 1 - x_1 = 1.$$

"The ESSENTIALS" of COMPUTER SCIENCE

Each book in the **Computer Science ESSENTIALS** series offers all essential information of the programming language and/or the subject it covers. It includes every important programming style, principle, concept and statement, and is designed to help students in preparing for exams and doing homework. The **Computer Science ESSENTIALS** are excellent supplements to any class text or course of study.

The **Computer Science ESSENTIALS** are complete and concise, with quick access to needed information. They also provide a handy reference source at all times. The **Computer Science ESSENTIALS** are prepared with REA's customary concern for high professional quality and student needs.

Available Titles Include:

BASIC
C Programming Language
COBOL I
COBOL II
Data Structures I
Data Structures II
Discrete Stuctures
FORTRAN
PASCAL I
PASCAL II
PL / 1 Programming Language

If you would like more information about any of these books,
complete the coupon below and return it to us or go to your local bookstore.

RESEARCH & EDUCATION ASSOCIATION
61 Ethel Road W. • Piscataway, New Jersey 08854
Phone: (908) 819-8880

Please send me more information about your Computer Science Essentials Books

Name _____

Address _____

City _____ State _____ Zip _____

Available at your local bookstore or order directly from us by sending in coupon below.